Android

Smartphones

SENIORS GUIDE

2023

A Step-by-Step Manual for the Non-Tech-Savvy to Master Your

Brand New Smartphone in 3 Hours or Less

John Halbert

Table of Contents

Introduction

A cell phone is like a portable computer that allows you to always communicate with dear ones via voice calls and text messages. Since the early 2000s, their popularity has skyrocketed, with 97% of Americans and 92% of people over the age of 65 now owning one.

With more people using mobile phones comes more and more sophisticated features, which may make it difficult to use, especially as a senior. The most intuitive features may be found in phones designed with the elderly in mind.

The convenience of cell phones may increase an elderly person's standard of living. For instance, they may facilitate an active social life by constantly communicating with loved ones through picture sharing and video conferencing.

It's no secret that phones are a huge step toward achieving personal freedom. As a senior, mobile applications are available that give you command over your health and fitness by maintaining a medical history record. Knowing how to use digital resources like the internet and navigational aids like GPS might be useful. In addition, you may always call for assistance using a mobile phone, which is a great safety feature.

The cognitive stimulation provided by games played on mobile devices, both online and offline, may help slow the onset of dementia. With your mobile phone, you can access many films and television series that can be watched on the go.

The Downsides of Using a Cell Phone

For the elderly, cell phones may provide some unique challenges. Think about the benefits and drawbacks carefully before making a purchase.

The Features May be a Challenge to Work With

You must be familiar with the features of a mobile phone before using one. It might be challenging to use if you aren't used to the technology. Time and guidance may be required before you can use your phone independently.

Simple to Steal

The odds of recovering a stolen mobile phone are low, with one in ten devices being taken each year and sold for profit. Anxiety is a common reaction to the thought of losing your phone.

Suiting Your Financial Constraints

The price of a brand-new smartphone might reach $2,000 and above. Even used or refurbished phones from a previous generation might cost several hundred dollars. Each month's payment must include the cost of the phone plus the cost of any applicable data, voice, or text plans. You may need to dig to find a phone and plan that works for your budget.

What To Consider Before Purchasing a New Cell Phone

It's easy to feel overwhelmed by all the choices available to you while shopping for a new phone, from the many phone models to the various service providers. Remember your available funds and the phone's capabilities when making these selections.

Choose a Mobile Phone Category

Flip phones, block phones, and smartphones are the three primary categories of mobile devices. A flip phone or block phone is ideal if you only want to make and receive phone calls and texts without accessing the internet. A smartphone has the greatest capabilities and can connect to the internet via cellular data or a Wi-Fi hotspot.

Select a Make and Model

Two companies produce most mobile phones and have some of the costliest options. However, there are cheaper options, such as older models with identical specs or ones from other companies. Another option is to purchase a reconditioned phone, albeit the quality of these devices may be questionable if purchased from an unknown vendor.

Discovering a Service Provider

In addition to the three big network providers, there are other smaller providers, each of which offers a unique set of monthly rates. Different tiers of these programs provide escalating perks like more data or a longer membership period.

Keeping Your Phone Safe

If you buy a smartphone, you should buy a case to protect it if it is dropped. Remember that phone cases are model-specific, so ensure you get the right one. Additionally, screen protectors provide an extra layer of protection against scratches, but they may reduce the responsiveness of smartphone touchscreens, making them less accessible.

Choosing the best mobile phone for seniors might take time and effort. With the right set of functions, you can maximize your phone's utility and learn to rely on it as a resource in your day-to-day life.

In the coming chapters, we will go over setting up your brand-new mobile phone, so you can have the best experience while using it.

Chapter 1: How to Turn On Your Phone and Setup Your Sim

Your excitement over utilizing your new Android-powered smartphone or tablet is palpable. Don't worry if the installation process seems difficult; let's walk you through each step.

Powering on your Android phone is the first step toward utilizing it. It may be preferable to get your device fixed if you suspect the Power button is damaged or the battery is dead. There are a handful of troubleshooting techniques you may attempt to turn it back on.

Today's smartphones often include setup wizards, making getting started quick and easy. You may have a somewhat different experience on various devices and Android versions. We use the official Android version, although they're almost identical in every other respect.

To power your smartphone, follow these steps.

- Ensure your phone is charged completely, then insert your SIM card and turn it on.
- Choose a language
- Connect to Wi-Fi
- Type in your Google login information.
- Choose a backup method and a method of payment
- Set date and time
- Install a security measure, such as a password or fingerprint.
- Setup up the Voice-Activated Assistant
- Get ready to start your app and media downloads

Here's a more detailed approach:

- Step 1

If your new gadget is a smartphone or tablet with mobile data capabilities, insert your SIM card. Insert the battery and snap on the back panel if it has one (this is becoming more unusual).

The power button on your new Android smartphone should be on the right side. Keep in mind that it might need charging before turning it on.

- Step 2

After turning on the device, the first thing you'll need to do is choose a language. Click Get Started after making your selection.

Sim card installation is optional but recommended during setup since you will be requested to do so if you haven't previously done it. Insert your SIM card or choose Skip to skip this.

You may be prompted to restore from another device (which will transfer over your apps and data) or set up as new when you first set up your smartphone.

- Step 3

If you want to use Wi-Fi, you will be prompted to join a network. Select your network from the list, enter your password, and click the Connect button.

- Step 4

You can sign in with your email address and password if you already have a Google account. If you don't, you can create one by clicking "Create a new account" and then continue to the next step. To continue, you must accept Google's Privacy and Terms of Service by clicking Accept.

- Step 5

During setup, you'll be asked if you want to enable features such as automatic device backups (recommended), using Google's location service to help apps determine your location, which will grant location access to specific apps when needed, and improving location accuracy by allowing the device to scan for Bluetooth and WiFi even when they're turned off (recommended, provided battery life isn't an issue). You can toggle each of these features on or off, depending on your preferences. The best feature is that you can change them anytime, so don't worry.

- Step 6

You might need to change the data and time zone if you purchased your phone from somewhere other than your current location. Knowing the right time is important for obvious reasons and to avoid any potential interference with your wireless network. You may also adjust the time-setting choices in your device's Settings if you aren't asked to do so.

- Step 7

Set up your fingerprint scanner immediately if your device supports it. You may either press add fingerprint or come back later to add one. You'll need to create a backup login method, such as a pattern, PIN, or password before you can set up a fingerprint. These will be explained further in the following sections.

- Step 8

Follow the on-screen directions to get started with Google Assistant if available on your phone.

- Step 9

In most cases, that will be the end of the required setup processes, although, as we said before, this may not be the case for all devices. A Samsung smartphone, for instance, can prompt you to check in with your Samsung account so you can access the Bixby assistant.

It's possible that after being introduced to your new gadget, you'll be given a quick rundown on how to alter the look and feel with tweaks to items like the wallpaper, widgets, and system preferences. This is important if there is one, but you may always just go right to the main menu if you like.

To see if any updates are available for your phone or its preloaded applications, launch the Google Play store app, tap the three lines symbol in the upper left, then My apps & games, then select Update All.

You can change your background and ringtone under the Settings menu, so it's worth exploring Android and becoming used to its features.

After completing these steps, you'll be ready to download applications from Google Play. To install an app like WhatsApp, just launch the app store, search for it, then tap the Install button. App installation will start automatically when you are requested to provide the necessary permissions.

How to Lock Your Android Device

You may be prompted to activate secure startup, necessitating entering a PIN whenever the phone is powered on. It's up to you to decide how secure you want your phone to be, but this feature can be bothersome. Select Require PIN to start the device or skip this step if you don't want to set one up, and then proceed.

Choose a memorable four-digit PIN and enter it. Avoid using obvious numbers like 1234 or 0000 and personal information like your birthday, which might be used to guess your PIN. When you are ready to go forward, use the right arrow key.

To be sure you didn't make a mistake, Google will have you re-enter your PIN and hit the arrow key.

A Fingerprint Setup for Android

Get ready to set up your fingerprint. Find the device's fingerprint reader. It is typically built into the Home button or situated just below the device's camera on the back. In most situations, the on-screen picture will have been adjusted to correspond to the place on your device, but if you're still unsure, you may check the user manual, review the product, or just look at it.

Keep your phone in its natural position, as the fingerprint scanner works best when your fingers are not forced to perform any kind of gymnastics. Then start tapping the sensor. Repeat this process many times, ideally with your finger in a slightly different position each time, so that a more accurate fingerprint scan may be created. You can go to the next stage as the on-screen image fills up.

Smartphones and tablets often have space for five fingerprints. If you like to use your thumb or index finger to dial, this feature might come in handy. Consider who else in your home could require fingerprint access if the gadget becomes popular. You may add the remaining fingerprints later if you want, but you don't have to input all five at once.

You can skip this step and piggyback on it much later. You can access this feature from your "settings" tab. This is what it should look like.

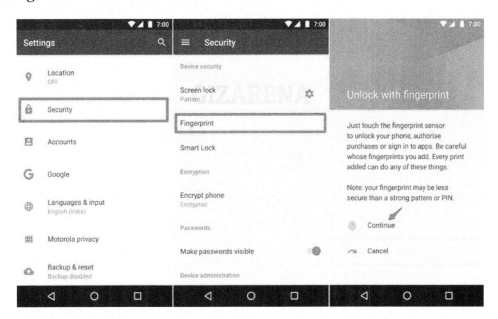

What to Do If Your Phone Won't Turn On

It can be terrifying if your phone stops working once you've spent a lot of money on it and it's no longer supported by warranty. It is nearly impossible to function in this day and age without a fully functional smartphone.

There's a strong temptation to write off your phone as dead and rush out and get a replacement, but we wouldn't recommend doing so. See whether your Android gadget is still alive by following these troubleshooting procedures.

Examine the Phone for External Damage

First, thoroughly inspect your phone. Have you recently dropped your phone? Is there any cracking or damage to the screen? Is there an expansion in the battery? (In this case, do not attempt to turn on the phone; instead, bring it to your local e-waste center.) Has there been any flooding? (you can verify this by inspecting the SIM card slot.) You may have to replace the part or get it serviced if there are hardware issues.

It is possible that your phone can fall, the glass can remain intact, but the screen may go blind. If that is the case, your phone may be functional but only has a broken screen. You might try restarting your phone by holding the power button for a few seconds. Test to see whether it vibrates at all when you turn it on.

There is also the option of using a virtual assistant such as Siri or Google Now. Make a call to your phone from another phone. A new display is required to repair a phone that still makes calls but has a blank screen.

Get the Batteries Back Up to Speed

Although it may seem absurd, it is also possible that the battery in your phone might be dead. If your phone's battery is dead, putting it into a charger may not immediately cause it to turn on. If it won't switch on immediately, try unplugging it and waiting 15 to 30 minutes.

If it doesn't work, the charger itself may be broken. Attempt recharging it from a different source, using a different cord and a separate portable power source. The charging port should also be checked since lint may easily become stuck in the pins and hinder charging. You may try charging the phone again after using a toothpick to remove debris from the port.

Force a Reset

If turning the phone off and back on by holding the power button doesn't work, you may need to resort to a hard reset.

The vast majority of Android phones appear and behave similarly. On many Samsung phones, for example, holding the Volume Up and Volume Down buttons and the power button activates the screen. Please consult the manual or go online for reset instructions if you have problems.

A phone with a detachable battery may be turned on properly by removing it, waiting a few seconds, and then replacing it.

Simply Return Your Device To Its Original Settings

There's always the possibility of resetting your phone to factory settings if nothing else works. (You should have a backup, right?) How this is accomplished varies by manufacturer, but here's a general guide to help you.

The method for accessing Recovery Mode on an Android device varies depending on the device. To navigate the recovery menu on a Samsung Galaxy S10e, hold the volume down, Bixby, and power buttons simultaneously for a few seconds. Hold the Volume Down and Power buttons to simultaneously access Recovery Mode on the Pixel 2.

To do a factory reset, enter Recovery Mode by holding the power and volume down buttons simultaneously. Similarly, if you can't locate this setting on your phone, try searching online for your model.

Reinstall the Original Software

If you still have trouble entering Recovery Mode after trying the preceding steps, you may need to enter a lower-level mode to reinstall the firmware.

Again, there is too much variation across Android devices for us to offer steps for flashing firmware on each one of them here. Installing the Android Debug Bridge on your computer is the first step towards updating the Pixel's firmware.

You'll need to research how to flash fresh firmware from scratch on your unique gadget, which may need its own specialized equipment. Warning: if you're not sure you can handle the complexity of these instructions, you should take your phone to a repair shop.

Get Yourself a New Cell Phone

It may be time to purchase a brand-new mobile device when all else fails.

Chapter 2: How to Setup Wi-Fi/4g/5g and Switch Between Any or All of Them at will

To check whether your SIM card is compatible with the Viva-MTS 4G network, send the *776# instruction to your phone or check the settings page.

When you use the USSD command to see if your card supports 4G, you'll get a message telling you whether it does. If your card is Viva-MTS 4G compatible, you'll have access to the best that today's technology offers.

How to Activate 4G/LTE on Android OS Devices

- Open "Settings"
- Tap on "More"
- Tap on "Mobile network."
- Tap on "Preferred network type."
- Choose "4G/3G/2G (auto)."

The steps should look this way (depending on the type of android phone you are using).

Chapter 3: How to Check Your Phone's Operating System and Update Your Phone

The Settings app on your Android smartphone is where you'll discover information about your device's Android version, security patch level, and Google Play system level. You will be notified as soon as new updates become available. Furthermore, you can see whether there have been any recent changes or updates.

To discover your phone's operating system:

- Determine your Android version.
- Launch the device's settings menu.
- Select About phone and then Android version from the menu that appears towards the bottom.
- Locate your device's "Android version," "Android security update," and "Build number."
- Upgrade to the most recent version of Android now.

- You should open the notification and choose the update option whenever you receive one.

Exercise patience as the update happens because it can take a few minutes to complete. Also, try not to shut down the app or simultaneously start any new activities with your smartphone. This is what your screen should look like.

If you have turned off notifications or gone without your device:

- Launch the phone's settings menu.
- Select System, followed by System update, from the bottom options.
- Your current status will be shown. Just do what it says on the screen.
- Download the latest Google Play system and security updates.

To find the latest available android version

- Tap the Google Security checkup icon to see if any updates have been released to address potential vulnerabilities.
- To see whether there is a newer version of the Google Play system, choose Google Play > Settings > About > System updates.
- Just do what the screen instructs you to do.

You may ask, when will Android updates become available?

- The device, manufacturer, and mobile provider might affect the time updates are released.
- Find out when you can expect updates for your android phone by contacting its maker or service provider for assistance.

Chapter 4: How to Setup Security Code for Your New Android

Password-protecting your mobile device is a good first line of protection against unauthorized data access. If your smartphone is lost or stolen, this will prevent unwanted people from accessing its contents.

You must enter your passcode every time you turn on or wake up your smartphone. If your device stores sensitive data, you should take additional precautions (such as enabling remote wipes) in addition to password protection.

Follow these instructions to establish a passcode on your Android device:

- From your device's applications menu, choose Settings.
- A menu will appear when you tap Security (or Security and Screen Lock), often under the Personal area.
- Click on the screen lock button. You'd usually find it under the Screen Security section.

- You have several options to choose from, and you can choose the lock type you want for your device.

- You may unlock your smartphone by drawing a pattern on a grid.

- PIN: To unlock your smartphone, enter a four-digit code.

- You may use a password that you feel confident using. There must be more than 4 characters in the password. Also, ensure that you choose a combination that isn't easy to guess.

The steps should look this way when you follow them.

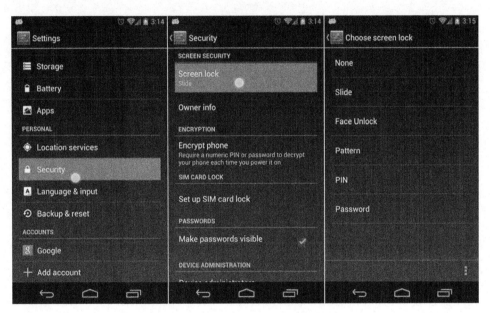

Chapter 5: How to Import Files from Old Android

Make sure all your data transfers securely when it's time to transition from your old Android phone to a new one. Fortunately, Google makes it amazingly simple to transfer all your messages, call logs, contacts, music, and images to your new phone.

Before switching to your new phone, you'll need to ensure that all the important data on your previous phone has been backed up to the cloud. When setting up your new phone, just sign in with your Google account to access Google Drive. Google automatically backs up your contacts, call logs, messages, and settings, including information about Do Not Disturb.

The best part is that you won't run out of space because Google automatically backs up data, which does not count against your 15GB free Google Drive storage allowance. This saves call history, app data, phone contacts, device settings, and text messages. Before switching to a new phone, check to see if this data is already in the cloud.

To back up your details to the Google Cloud, follow these directions:

1. Access Settings from your applications or the Quick Settings menu.

2. Navigate to the page's bottom by scrolling down.

3. Select System from the menu.

4. Select Backup.

5. Verify that the Backup to Google Drive toggle is turned on.

6. Press Backup to sync your phone's most recent data with Google Drive.

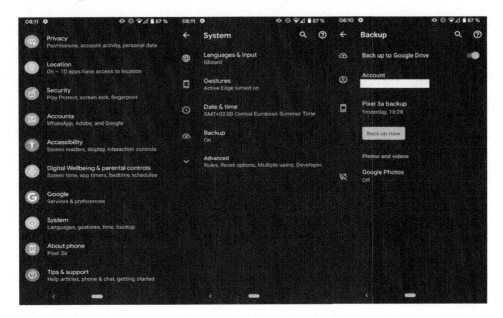

Check contacts.google.com to determine whether all of the contacts on your old phone have been transferred to your Google account. Your new phone will only have the contacts listed on this page.

After the data has been backed up, you may begin configuring your new phone. However, check whether Google Photos is backing up your images and videos first.

How to Get Your New Android Phone To Accept Images and Videos

Google Photos is a fantastic tool for archiving images and movies. It smoothly backs up data, provides flexible sharing restrictions, and intelligently groups face using on-device machine learning. Your images and videos may be saved at their original quality, the quality they were originally captured. You can choose Storage saver to keep the same degree of detail but condenses the file size.

This is a fantastic alternative because Photos no longer allows limitless uploads. Once your storage is full, you must purchase a Google One subscription. It is the finest method for backing up your Android phone's photographs and movies. You'll be able to view your photographs and movies from your new phone and the internet after the data has been backed up to the cloud. Here's how to get started with Photos if you haven't already.

1. Start your phone's Google Photos app.
2. In the top-right corner of the page, tap on your profile photo.
3. Pick the Photo settings option.
4. Choose Sync & Backup.
5. Verify that the Backup and sync toggle is turned on.
6. Ensure that High quality is selected for the Backup mode.

That is all you need to do. After your images and videos have been securely backed up to Google Photos, you may begin configuring your new phone.

How to Transfer Data from Your Old Phone to the New One

You'll need your old phone to transfer data, so don't wipe the data yet. You can move your old SIM card to the new phone. Let's get started!

1. On the welcome screen, choose your device's language and press Start.
2. Select and sign in to your home Wi-Fi network.
3. Select Next to start transferring data from your previous phone at the Copy applications & data page.
4. Connect your old phone to your new one with a USB-C to USB-C cable, then press the Next button on the new device.
5. Next, attach your new phone to the other Type-C port.
6. Click Next once again.

Choose the files and data you want to transfer and follow the on-screen instructions. Because Android has become so user-friendly, it also logs into your Google account and sends all your saved WiFi credentials. Your data is safely transferred to the new device using a direct data transfer technique, eliminating the need to enter your Google account password or two-factor authentication.

You'll also notice the Play Store operating in the background and recovering data from your previous phone after the setup, and you reach the home screen. You may wait for it to finish while setting up the rest of your new phone's settings. This process might take up to an hour, depending on the volume of data loaded on your previous phone and your internet connection.

You can recover your files and data from your cloud backups if anything copies improperly. However, this approach often works exceptionally well.

Chapter 6: How to Set Up an Existing Account or Create a New One with Google

You could be seeking a means to establish a Google account, whether you just purchased your first Android device or simply want a new method of communication. The procedure may seem complicated, but it's rather straightforward. Here's a simple method for creating a Google account.

Because most of Google's services, such as Gmail, Docs, Drive, and Photos, are free, entering a credit card is optional, and opting out will have no consequences.

What, then, would motivate you to include a credit card? It may be used to pay for other Google services, such as YouTube Music subscriptions and Play Store purchases of applications and games. To restore your account if it is lost or stolen, Google will also need your phone number.

How to Setup a New Google Account on Your Android Phone

You may have an unlimited number of Google accounts, and it simply takes a few minutes to create a new one. Select the Accounts option in the Settings app on your Android smartphone after grabbing it. Next, choose Google by tapping Add account at the bottom.

A screen where you may sign in to your account or make a new one will display. Choose the option to "Create an account," then adhere to the on-screen directions to input your data (including a phone number), choose a username, and accept the terms of service to finish the process.

Follow these step-by-step instructions to create a new Google account now:

- Go to your device's settings.
- Choose Accounts.
- Click Add account.
- Choose Google.
- Pick Make an account.
- Enter your personal information, choose a username, and other instructions as directed on the page.
- To establish a Google account, click the "I Agree" button.

How to Create a Google Account on a Browser

Not a fan of the phone method? Using a computer and online browser to fill out forms and establish an account may be quicker and simpler.

This whole process can easily be done online.

This is how to create a Google Account in a browser, step by step:

- Visit the Sign Up page at accounts.google.com.
- Follow the prompts on your screen to enter your name, username, and preferred password.
- Choose Next.
- Enter your gender, birth date, recovery email, and phone number.
- Choose Next.
- You may now confirm any phone numbers you supplied.
- Observe further directions.

Chapter 7: How to Charge Your Phone at Home or with a PC Cable

You should use the wall charger that came with the phone; however, if you do not have a charger, you can use a USB cable to connect it to a computer. A PC's maximum USB 3.0 current is 0.9A, implying that phones charge slowly.

You may charge your phones via USB cables connected to your personal computers because it is quick and easy. It's common for people to use this strategy in an emergency, especially in public places with limited power outlets, such as airports. Nonetheless, the phone's power consumption appears higher than it should be after charging. Then they'll consider whether using a computer as a phone charger is a good idea. So, where do these circumstances come from?

What are the Risks of Using a Computer to Charge Your Phone?

Charging a smartphone over a USB from a computer is risk-free and won't harm the battery. This is because a single USB cable may be used for wall and computer charging. Only the charging current is different in this mode. Regular chargers and computers are only differentiated by the rate they charge.

- May Damage the Phone's Battery

Because the voltage of the computer's USB port is less than that of the charger, charging a mobile phone battery via a USB port on a computer may cause damage to the battery. Furthermore, the current and voltage will fluctuate due to the computer's fluctuating power requirements. As a result of the electric ion damage, the mobile phone's battery life will be reduced. For example, energy consumption on a computer during video playback differs from that during standby.

It is generally not a good idea to use a computer to power your phone's battery since the activity of lithium batteries may be affected by the prolonged use of a computer to charge mobile phones. There is a need for a specific charging voltage due to the prevalence of lithium batteries in modern mobile phones. The USB connection on your computer supplies a 5V/0.5A current, but the lithium battery requires far more voltage and current to be fully operational. As much as possible, try using a charger connected to a direct power source whenever you want to charge your smartphone.

- Poor Lithium Battery Performance

Even though people have varying viewpoints, there is still much to be gained through listening to them.

Why does the phone's battery die so quickly after being charged in a computer?

Because the phone isn't completely charged, it loses its energy quickly after being charged through a computer. In other words, even if the phone says it is fully charged, that's just a representation of the virtual power it is drawing. In truth, it is related to how a mobile phone's battery is charged. There are three phases to charging a mobile device: quick charging, continuous charging, and trickle charging. Even if the system's battery power indicator reads 100% after the first two phases, the battery does not achieve the true saturation condition. A trickle of electricity will have to flow through it to charge it completely.

- Trickle charging

Since batteries gradually lose capacity due to self-discharge after being charged, trickle charging is used to compensate for this. After a full charge, the battery will lose around 5% of its nominal capacity due to self-discharge. A trickle charge from the computer is now powering the phone to make up for the self-discharge it has experienced.

When we see a full battery indicator, we assume the charging process is over. Your phone is still using inefficient trickle-up charging. Even while your phone says it has 100% charge, you may only be getting 95%, which is why you may think charging over a computer's USB is inefficient.

- Virtual and Actual Power

Please wait a few minutes before removing the phone from the USB port of your computer after charging. Charge it for an additional half an hour if you have the time. An option is to upgrade the phone's battery capacity. Another function is to keep the battery's components safe, extending its useful life.

Chapter 8: Learning About Your Phone's Main Functions (Send and Receive Calls, Messages, Videocalls, Gmail, etc.)

You may place calls from the Phone app and other applications or widgets that display your contacts.

Usually, you can touch a phone number wherever you see it to make a call. Google Chrome may allow you to touch italic phone numbers to copy them to the dial pad.

To start making calls, download the Phone app from the Play Store if you don't already have it.

It's possible that your device isn't supported if you can't download the Phone app.

To make the app your phone's default app after downloading it, follow the on-screen instructions. Send a text or call. You must agree to the request to make the phone app your default to use it. After downloading, you can launch the Phone app on your phone.

Choose who to call:

- Tap Dialpad to input a phone number.
- Select a stored contact by tapping Contacts.
- Tap Recents to choose a number from those you've just recently phoned.
- Select a contact from the Favorites list by tapping Favorites.
- Call by pressing the button.
- Tap End call after you are done with the call. If your call has been minimized, drag the call bubble to the bottom right corner of the screen.

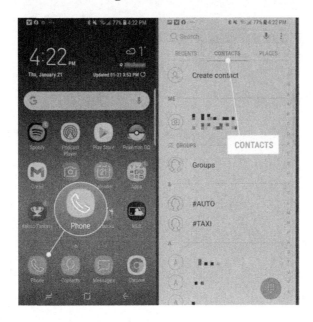

You can also use select compatible carriers and devices to make video calls, conference calls, or RTT calls.

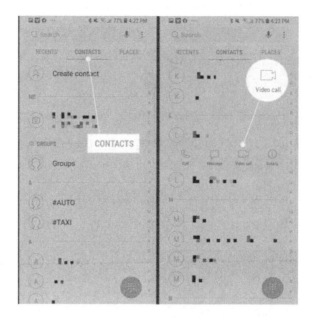

Whether to accept or decline a call

- The screen displays the caller's number, name, or ID information when you receive a call. If Google can verify the number, you'll see Verified next to the caller's name or phone number.
- When your phone is locked, touch Answer or slide the white circle to the top of the screen to answer the call.
- If your phone is locked, slide the white circle to the bottom of the screen to reject the call or press Dismiss. Callers that are disconnected may leave a message.
- Swipe up from the New message symbol to decline the call and text the caller instead.
- Your current call is put on hold when you take a call while on another.

If Google Assistant is activated, you may answer or reject a call using your voice. To use this feature, you can use these voice commands:

- "Hey, Google, pick up the phone."
- "Google, ignore this call."

During an active call:

- Dialpad may be tapped to display the keypad.
- Tap Speaker to switch between the earpiece, speakerphone, and any connected Bluetooth headset.

- Tap Mute to turn your microphone on or off.
- Press the "hold" button to end a call without hanging up. Tap Hold once more to pick up the call once again.

Seniors-Friendly Video Calling Device

When you are unable to visit your elderly loved ones in person, video conferencing is an excellent substitute. FaceTime, Skype, and the increasingly popular Zoom video conferencing service are just a few excellent online video chat options available today.

On the other hand, not all interfaces are naturally arranged for use by less tech-savvy older people. Here are some simple video communication gadgets that may assist the elderly in keeping in touch with their family and friends via the convenience of video calls.

GrandPad

The GrandPad is a safe and user-friendly tablet that facilitates communication between grandparents and their loved ones in the modern world. The elderly person may quickly and simply look at pictures and watch videos of the family, play games, check the weather for themselves and their loved ones, listen to music, read emails, talk on the phone, and even engage in a video chat.

The GrandPad stands out because of how user-friendly it is for the elderly. All the capabilities, including video chatting, are designed with the user's age in mind.

ViewClix

ViewClix is great for families who are always on the road but still want to stay in contact with their loved ones, and it's also great for elders who may be technologically inexperienced.

ViewClix makes it simple for loved ones to connect via sharing photos, transmitting live video, and affixing Sticky Notes to the ViewClix photo frame.

Video chat options in ViewClix include:

- Using the free ViewClix mobile software on your Android smartphone or tablet, you may quickly and easily make video calls to the frame. ViewClix video calls are also compatible with Mac and PC browsers.
- In auto-answer mode, a call may be answered automatically, eliminating the need for a senior to manually answer calls.
- A smartphone or tablet's rear-facing camera may be used for a "flip cast," allowing you to show your elderly loved one a live video feed of the family.
- Safe and sound, only approved family members may make video calls.
- Each ViewClix frame has a high-quality external microphone designed for conferences.

Videophone Konnekt

The Konnekt Videophone is a simple telephone modified for the elderly, disabled, and hearing-impaired (Konnekt also offers to caption on their device). With a single tap, you can initiate a video call to any Skype user on any computer, tablet, smartphone, or mobile phone.

Face-to-face communication with loved ones is now easier than ever with the Konnekt Videophone's intuitive ONE-TOUCH-TO-DIAL interface. No prior knowledge of computers is needed. Not a single menu, no login, and no frills.

Suppose you have trouble using a traditional telephone because of vision or hearing impairment, shaky hands, inability to move, cognitive impairment, or dementia. In that case, the Videophone may be the solution you've been looking for. Through the Videophone, Konnekt's consumers may join in on the digital communication era without the learning curve often associated with such technology.

Google Nest Hub + Duo

Making and receiving calls from others using Google Duo on the Google Nest Hub is possible. All your devices, whether they run Android or iOS, can use Google Duo, including laptops and Smart Displays. A great conversation starter is the Knock Knock function, which provides a live video preview of the person contacting you.

The Google Nest Hub has a lot of extra functions that may be overkill for less tech-savvy elders, but it has fantastic support, is backed by the Google brand, and has a straightforward UI.

Facebook Portal

The Facebook Portal is an intelligent video-calling gadget that can be used with either Facebook Messenger or WhatsApp, making it ideal for keeping in touch with elderly family members who use Facebook. Features that stand out the most are:

- Chat with loved ones in a snap-through video using your existing Messenger or WhatsApp account, even if they don't have a Portal.
- Don't miss a beat! You may walk about and converse freely while the smart camera keeps everyone in view.
- Listen and speak up. Smart Sound is a useful tool for public speaking by boosting your voice and dampening ambient noise.
- Use a single toggle to turn off the camera and microphone or cover the lens. Video chats made using the Portal are secure.
- Read along with your kids' favorite stories while being transformed into characters via music, animation, and augmented reality.

Amazon Echo Show

Others who have the Alexa app or any Echo device with a screen may make video calls to those who have the Echo Show.

The Echo Show is similar to the Google Nest Hub in that it offers a wealth of capabilities for older people who are somewhat computer savvy.

There are several functions available on the Echo Show.

- A high-definition 8-inch display with dual speakers
- Ask Alexa to play movie trailers, TV programs, movies, or the news, and prepare to be entertained. You may also tune in to radio shows, podcasts, or audiobooks.

- Command your high-tech dwelling - You may command compatible gadgets with your voice or through a straightforward interface. You can also use Alexa to turn on/off lights and regulate temperatures.

- Transform it into something you own - Exhibit your Amazon Photos collections. The home screen may be personalized. Develop daily rituals to ease into the day.

- Tailor-made for the way you live - Follow along with the recipes while you cook. Maintaining schedules and to-do lists is a breeze. Check the forecast and traffic conditions before leaving the house.

- Developed with your privacy in mind - Pressing the button once will electronically turn off the camera and microphones. The camera may be covered by sliding the internal shutter.

Chapter 9: How to Setup Play Store, Credit Card, Download, Update, and Uninstall an App

You must link a credit or debit card to your Google Play Store account to download premium applications and purchase subscriptions. Following the instructions on your screen, once you open the app, you can add a new credit card to your Google Play Store account, update an existing card, or remove an expired one.

To successfully add a new card to your account, the card you're trying to add must have some balance. To validate it, Google will have to do a short reversible transaction.

The default applications on an Android phone are rather versatile. Of course, you can get it to do much more with the plethora of applications available, which range from providing a weather prediction to helping you organize your life, with games thrown in for good measure.

How to Download Apps from the Google Play Store

- Open the Google Play app on your phone.
- Find an app by searching using the search bar at the top of the screen.
- Tap Install

Google's Play Store makes it simple to obtain useful apps. This app comes pre-installed on almost all Android devices as a piece of the suite of applications that Google bundles with Android. The app's icon is a slanted multicolored triangle superimposed on a white background to locate Play Store. For the sake of clarity, this is how it appears.

You'll be taken directly to the Play Store if you click the icon. In addition to the applications shown in the main section of the screen, a row of icons at the bottom of the screen will categorize the available apps into four categories: Games, Apps, Movies, and Books.

Select the category that should contain the app you require, and then, if you know the app's name, enter it into the top of the screen where the search box is. It's now as simple as typing the app's name into the search bar; ideally, it will come up first. If you click it, the app's website will open, and you can simply click the Install button to begin the download.

Remember that certain applications may cost money, while others may be free. If you choose one that requires payment, you'll be asked to add a payment method; however, you may also use Google Play gift cards if you don't want to use a credit or debit card.

How to Install Apps on an Android Device

Above the highlighted applications, you'll see choices labeled "For you," "Top charts," and "Categories," which you may use if you're still unsure about which one to download. These will show you a list of the most downloaded ones from the app store, and you may install any of them with a single touch.

Although this is not the official method, it is among the most secure and user-friendly ways to get Android applications. There are many other apps that you can access by downloading them directly from different websites. However, the main advantage of getting apps from the play store is that you don't have to be afraid of getting attacked by viruses or malware because of the safety and security measures taken to ensure that the apps on the play store are free from these. However, if you risk it, download and install apps from other sources.

However, it should be noted that while Amazon and, as far as we can tell, the App Gallery are reputable stores, many others may be home to dubious programs or viruses due to the more open nature of their catalog. As a result, we recommend staying within the confines of the Google Play Store; however, if you insist on venturing outside, arm your Android device with a solid antivirus app.

Another option is to download the software as a .apk file from a website, potentially even the app's creator, just as you would for many Windows applications. Antivirus software is recommended if you want to perform anything like this since, once again, there is a chance that it might contain malware.

You'll need to change your device's settings so you can download applications from places other than Google Play before installing any third-party apps. Even though the exact names of the following features may change somewhat depending on the device, you're using, their general nature remains the same. To accomplish this, launch the Settings app, then go to Apps by tapping the cog icon in the upper right. Choose "Special access" from the list. Select Install unknown applications at the bottom of the following page.

There will now be a list of available applications. Apps may be installed from any of them, so choose the ones you're most likely to use (Chrome should be at the top of your list) and switch on the Allow from this source setting. This concludes the discussion. You may now access those other app shops and download the programs directly onto your phone. Make sure you keep them up-to-date regularly to ensure the highest level of security.

How to Add a New Card to Your Google Play Account

Adding a new credit card to your account is simple and takes only a few minutes. Follow these steps to finish the process right away:

- While the Google Play Store is open on your device, tap your profile photo in the upper right corner.
- Go to Subscriptions and payments.
- Select Payment Methods by tapping it. If you have already added cards before, you will see a list of them when you tap this button.
- After entering all the necessary card information, select Add a credit card or debit card and select Save.
- To authorize your card, Google will carry out a little reversible transaction.

Voila! You can now purchase apps on the play store using this card. Feel free to add more cards if you wish.

How to Take a Current Card Off Google Play Store

Your card may need to be removed if it has expired or if you want to use a new one. Follow these steps to completely remove a card from the play store.

- Tap your profile image in the top-right area of the app after opening it.
- Select Payment Methods from the Payments and Subscriptions menu.
- This will reroute you to pay.google.com in your browser, where, if requested, you may sign in using your Google account.
- You may remove a card from the Google Play Store by selecting Remove.
- Without taking out the card, hit Alter, make the required adjustments, and tap Update to update or edit your card information.

Purchasing apps on the play store with your card is just easy. Once your card is set up, you may rapidly make purchases and, if you have numerous alternatives, even switch between various cards while making a transaction. When making a purchase, you may also add a new card by visiting the details page for a bought item and following the onscreen instructions. Your account will automatically store this card for future use.

Chapter 10: How to Change the Background and Lock Screen Photos on Your Android Phone

The option to customize smartphones is one of their nicest features. Compared to the days of the traditional flip phone, when the trendiest phone available was a Razor that came in vibrant colors, we now have many more options regarding how your phone looks inside and out.

You may now modify the outside of your phone and how it functions. Some claim that your phone's lock screen resembles a modern-day locket since many people use it to display images of loved ones, including pets and humans, so they can see them again throughout the day.

It's simple to change your lock screen at any moment, whether it's a photo of someone you love, a scene you genuinely enjoy, or something completely different. There are two methods to update your lock screen if you have an Android phone. You can use a picture you've shot or one of Android's pre-installed themes.

Here's how to modify Android's lock screen.

1. Launch the pictures app and choose the image you want to utilize.
2. Click on the three dots in the upper right corner to access the options menu.
3. Locate the "Use as" option by scrolling to the right of the list and tapping it
4. Select the "Photos Wallpaper" option from the pop-up menu.
5. Modify the image's size and zoom level as desired, then click "Set Wallpaper" in the upper right corner.
6. Select "Lock Screen" from the menu or "Home screen and lock screen" if you want it to be the background of your phone after it has been unlocked.
7. Conversely, you can go in from the settings app. Launch the app, click display, select lock screen display, and select the photo you want to use. It is that simple.

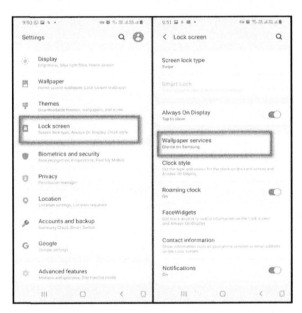

Would you like to make a default wallpaper your android lock screen image?

Follow these steps.

1. Launch the Settings app on your smartphone.
2. Pick "Display" from the options menu.
3. Choose "Wallpaper" under "Display" in the menu.
4. From the list, choose a category to peruse through in search of your new wallpaper. Search until you locate a picture you like.

To use one of your photographs as your backdrop, you may also select "My Photos" at this point. Afterward, zoom and crop the image you have selected to fit your screen size, and you are good to go.

Chapter 11: How to Set Up and Use Google Services/Apps

Google services are a great way to stay organized, keep up with friends and family, and keep track of all your favorite places. Google offers a wide range of services and apps that can be extremely helpful in both our personal and professional lives. We will cover how to set up and use some of the most popular ones.

How to Set Up and Use Google

How to Set Up a Google Account

You can sign up for a Google account by visiting https://accounts.google.com/signup in your browser. If you already have a Gmail address, you'll be asked to verify it before proceeding with the registration process. If not, click "I don't want to use my current email address" and enter an address that suits your needs best (you'll be able to change this later).

Once logged into your new account, check under Settings > Mail & Contacts > Add-ons/Gmail and select whether or not you want to access all of Google's services (like Gmail) from within this one app instead of having them spread across different ones like Yahoo Mail or Outlook Express—the default setting should work fine for most people though!

How to Use Gmail

To create a Gmail account, go to the top-right corner of your screen and select "Sign in to Google." Insert your email address and password next. If you already have an existing Google account with another service (for example, YouTube), this process will differ slightly depending on which one you're using.

Once everything is set up and ready for use, sign back into Gmail by clicking on its logo at the top of any page or from within an email message—this is also where you'll find options like managing labels and filters (more on those later).

How to Use Google Calendar and Google Meet

To create an event, tap the "+" sign in the lower-right corner of your calendar. Then choose "New Event," enter a title and description for this event, and select who should be invited to it (you can add multiple people at once).

Once you've added all of your attendees' names, tap "Save." You'll receive a reminder notification when there's time left before the event starts.

To view other calendars on Google services/apps—including Google Calendar—tap More at the bottom right corner of any screen where options are available such as sharing an image with someone else via email or text message, setting reminders, accessing settings, etc.

How to use Google Drive

You can store all kinds of files in Drive—like videos, images, and documents—and access them from any device with an internet connection. To use Drive, you'll need to create a Google account and sign in. Once signed in, you can create new folders and upload files from your computer or mobile device.

- Create a folder for your project.
- Share files with other people. You can share files with anyone with the same Google account by emailing them or sending them a link from your device.
- Backup your files. You can back up all of the data in your Google Drive account anytime by going to the Settings tab and selecting "Back Up Now." The backup will be stored locally on your computer, which makes it easy to restore if something

happens to make it unreadable or inaccessible through other means (like losing access at work).

To get started with Google Drive, open your app launcher and locate the drive app. Tap to launch the app and follow the prompts on your screen to get started.

How to Use the Chrome Browser

Google Chrome is the most popular browser in the world. It's available on Windows, Mac, and Linux computers, making it an excellent option for those who want a cross-platform experience.

Chrome has a built-in ad blocker that can turn on or off, depending on your preferences. The ad blocker works by blocking ads before they even load on your browser, so you won't have to deal with them when browsing online (you'll still see them if you click through). This feature also prevents trackers from tracking your browsing habits and other malware infections—a nice bonus!

To use Chrome browser, simply:

- Ensure that you have logged into your Google account using the steps we have shown you.
- Go to your app launcher and find the Chrome icon. Tap to launch it, then follow the steps on your screen to accept the terms and conditions.

- To browse the internet, click the search bar (the big bar at the top or button of your screen, depending on your android phone model), type in the website you want to visit and browse away.

Remember that you must be connected to the internet before you can browse.

Google Docs is a free cloud-based word processor and spreadsheet program available online. This implies that you can use this app to write on your smartphone. With it, you can take notes, write social media posts, or perform other writing tasks. The best part is that you can access Google Docs for free once you have created your Google account.

Google Docs has many features and functionality that you can use to improve your work productivity. Some of these include:

- Collaboration between multiple people working on the same document (everyone sees their edits)

- The ability to add comments in line with each other's text (this helps keep everyone in sync)

- An "AutoCorrect" feature that automatically fixes typos made by multiple people working together on one document

To use Google Docs, simply open your app launcher, locate the Google Docs app (if it is already installed on your device, or download the app from the play store), then open the app. Once it is open, click the blue "+" button at the bottom right side of your screen to open a new document.

You are good to go.

How to Use Google Lens

Google Lens is a camera app that allows you to take pictures and videos, recognize text, and perform other tasks using the camera on your phone or laptop. This helpful app allows you to identify objects with your camera or saved pictures alone. The first time you launch Google Lens, it will prompt you to set up some settings related to privacy and accessibility.

- Open your camera app (or any app that supports taking photos).
- Tap on the 'Settings' button in the top right corner of your screen. This will open up another menu with information about each feature available through this particular mode of operation; tap "Google" at the top left if it isn't already selected for viewing purposes alone.
- At the bottom, find and tap on discover. Then take and upload a photo you want to use for your search. Once you click the submit button, your results will pop up on your screen.

In summary, Google is a search engine and offers many other services. These include:

- Gmail - This email service comes with your account and lets you send emails to others. You can set up multiple accounts if you want to.
- Drive - If you use Google Drive regularly, this will be where most of your documents are stored. Using this service, you can also share files with friends or collaborate with colleagues!

Google has a lot of excellent services and tools that can help you manage and explore your new smartphone's powers. If you're new to using them, it might be hard to know where to start. Try out all these tools as soon as possible to become second nature for you when working on projects in the future.

Chapter 12: Manage, Add, Delete, Import, and Send Contacts To Others

Here's another interesting fact about your smartphone. You can save your contacts, emails, phone numbers, home address, and so on in Google contacts. Every contact saved to your Google account will automatically synchronize with your Google contacts. This will come in handy as you can always access the relevant information from any connected device or avoid the risk of losing your relevant contact details if you lose your phone in the future.

How to Save Your Contacts

Locate the contacts application on your Android phone and click on it.

Look to the bottom right on your Android phone, and you'll see a plus sign. Click on it to add your contacts.

To add a contact, here are the steps to follow.

- Enter the information required for the contact you want to save, such as name (First name, last name), phone number, and email.

- Select the Google account where you intend to save the contact (this applies if you set up more than one Google account on your smartphone). Locate your preferred email account and click on the arrow that goes down.

- You might also want to include additional name details. To do this, click on the arrow next to "Name."

- Also, to include a photo of the contact you're saving, look at the top of your screen and click on Add contact photo. This step is optional.

- To include more information about the contact, such as city, state, neighborhood, and so on, click on more fields. This is another optional step.

- After entering all the necessary information, you can save the contact by clicking Save.

This is what your screen should look like.

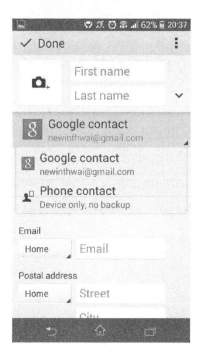

How to Import Contacts

You can include all your contacts in a Google Account. Even after you have imported your contacts to your preferred Google account, the contacts remain in your previous account.

How to Import Contacts from a SIM Card

If you have saved your contacts on your SIM card, you can import them to your Google account. How?

- Ensure your SIM card is inserted into your Android phone.
- Locate the contacts application on your phone and click on it.
- Look to the bottom, click on Fix and manage Import from SIM.
- If you have more than one account on your phone, select your preferred account where you want the contacts to be imported.

How to Import Contacts from a VCF File

First things first. A VCF (Virtual Contact File) is a type of file that can store contacts so multiple users can easily access them. If your contacts aren't saved on your SIM but saved to a VCF file, you can transfer them to your Google account. Here's how:

- Locate the contacts app on your Android phone.
- Look to the bottom, click on Fix, and manage Import from file.
- If you have more than one account on your phone, choose the preferred account where you want your contacts to be saved.
- Choose the VCF file you want to import and allow the process to be completed.

How to Move a Contact

Apart from importing a contact to your Google account, you can also move a contact. Even though they sound the same, they're a bit different. While inserting a contact into your Google account keeps the contact intact in both the new and former accounts, moving a contact deletes it from the initial account.

Follow these steps to move contacts in your new smartphone.

- Locate the contacts app on your Android phone.
- Choose the contact you want to move.
- Look to the top right, click Menu, More, and Move to another account.
- Select your preferred account to where you want the contact to be moved.

How to Delete Contacts

- Locate the contacts app on your Android phone
- If you want to delete a single contact, click on the contact. Look to the top right, click on more, and then Delete.
- Suppose you want to delete more than one contact; hold down on the first contact you want to delete for a few seconds. This will highlight the first contact. Then, select other contacts you want to delete at once.
- Click on delete and then delete.
- If you want to delete all your contacts at once, look to the top right. Click on More and then choose All, then you can delete.

Please note the contacts you delete are moved to the Trash on your Android phone. Then after 30 days, they are deleted permanently. If you don't want to wait for the time-lapse, but want to delete the contacts in your trash, then:

- To delete a single contact in your trash, Click on the contact and then delete it forever.
- To delete more than one contact: Click on one contact for a while, click on the other contacts you wish to delete and then delete forever.
- To delete all your contacts, Click on empty trash, and they are deleted forever.

Chapter 13: How to Setup Your Camera and Get the Best Photos and Videos

Another major advantage of having a smartphone is that most of them come with decent cameras that you can use to take amazing pictures and videos on the go. Imagine how awesome it would be if you could easily capture the interesting moments of your day and save them automatically to your android phone.

The camera can easily become one of your favorite features of your new android phone, and you can also explore a new career path in photography (even if you are just using it as a hobby).

Follow these steps to set up and start using the camera of your new android phone.

How to Launch Your Camera

- Interestingly, setting up your camera doesn't involve complicated steps. It is as easy as locating the camera app on your new smartphone and launching it. Once you open your camera app, you should see something like this.

- Clicking the settings button (the one circled in red above) will open you up to the settings page of your camera app. You can activate or deactivate the shutter sound, activate quick launch (where you can open your camera by just pressing the power button twice), and change your storage location. You can activate many other settings on this page, but the functionalities will differ depending on your Android phone.

- Taking photos is easy. Once you have launched the app, turn your camera toward the person or thing you want to capture and ensure the image is well positioned on your screen. Most times, your android phone will try to resolve the image you see, as some grid lines will automatically appear on the screen. Allow for some seconds to pass while this happens. Any picture you take while this resolution is going on will be blurry.

- Take your photo by clicking the big white button at the center of the screen. The shutter button may also be colored red, depending on your device's make.

- You can also take a picture of your face using the "selfie" option. To do this, simply click the toggle button anywhere on your screen. This button usually looks like 2 arrows in the shape of half-circles, pointing toward one another. Explore this option when you want to get uninterrupted shots of your face by yourself.

How to Make a Video

- Locate "video" on your screen with your camera app still open. Ost times, you will find this around the same place where you have other words like "picture," "portrait," etc.
- Click on "video," which will open up your video interface. When it opens, lick the big white button toward the bottom of your screen and have fun shooting your video.
- Some things that make a good video/photo include the angles and lighting. When taking photos, ensure there is adequate lighting, preferably natural lighting. Without natural lighting, you can compensate with your phone's flash. To activate the flash, click on the button that looks like "lightning" (the zigzag lines) on your screen.

For the sake of reference, the flash button is the one that is next to the settings button in the last photo you saw in this chapter.

Again, never forget the role of perfect angles while shooting photos and videos. You may have to experiment with standing at different points before you can get the perfect angle and lighting.

How to Import Photos, Export, Send, Receive, and Set Up a Gallery

Have you ever wondered how to import photos? You may have seen the option in your phone's settings or an app and weren't quite sure what it meant. You may have even tried it out, only to be met with a confusing array of options and not knowing where to start. We will guide you through importing photos from your phone to your computer. We'll also cover exporting photos, so you can share them with others and send and receive photos. Plus, we'll show you how to set up a photo gallery so that you can beautifully view your pictures.

How to Open the Gallery

Assuming you have the stock Android gallery app:

- Tap the Gallery icon to open the app.
- By default, the Recent tab is selected. This chronologically displays all the photos and videos, starting with the most recent.
- To view photos and videos stored in different albums, tap the Albums tab.
- To view only certain media types, tap one of the following tabs: Photos, Videos, or Stories.

How to Find Your Photos

Assuming you've already taken some photos with your Android phone, the next step is getting them onto your computer so you can do something with them. The good news is that this process is pretty straightforward, and there are a few different ways to go about it.

One of the easiest methods is to connect your phone to your computer via a USB cable. Once connected, open up your computer's File Explorer (or Finder on a Mac) and look for a new drive or folder labeled "My Phone" or something similar. Inside, you should see all the folders where your phone stores photos and other files.

If you want to access your photos from anywhere, you can upload them to a cloud storage service like Google Drive, Dropbox, or Microsoft OneDrive. Most of these services also offer mobile apps to view and download your photos on any device with an Internet connection.

How to Import Your Photos

If you've already taken some photos with your camera, the next step is to import them onto your other editing apps so you can start working on them. Follow these steps to import your photos and start working on them immediately.

- Connect your phone to your computer using the appropriate USB cable. If you're using a DSLR, this will usually be a USB cable.

- Turn on your camera.

- Open up the photo editing software of your choice. We recommend Adobe Photoshop Lightroom, but many other options exist.

- In Lightroom, go to File> Import from Device > choose your camera from the list.

- Select the photos you want to import and click Import.

- That's it! Your photos are now imported and ready for you to start working on them.

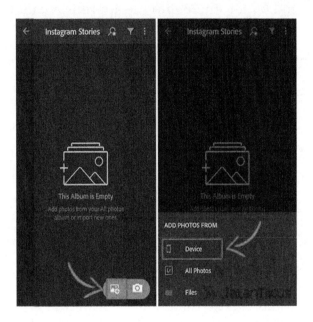

How to Export Your Photos

When you want to move pictures from your Windows computer to your Android device, there are several ways to do it. You can either use a USB cable to connect the two devices or export the photos from your computer and then import them onto your Android device.

If you're using a USB cable to transfer pictures, ensure a suitable cable. A micro-USB cable will work for most Android devices, but if you have a newer phone or tablet, you may need a USB-C cable. Once you have the right cable, connect your Android device to your computer.

On your computer, open File Explorer and find the photos you want to transfer. Once you've found them, select them all and copy them. Then, go to your Android device and paste them into the appropriate folder.

If you want to export photos from your computer to import them onto your Android device later, open up File Explorer on your computer and find the photos you want to transfer. Right-click on one of the selected photos and choose Export.

Choose where you want to save the exported File, then click Save.

Repeat this process for each photo that you want to export. Once they're all exported, go to your Android device and import them into the appropriate folder.

How to Send and Receive Your Photos

Sending and receiving photos is super easy once you have set up your gallery and started taking photos and videos with your smartphone. Select the photo from your gallery and click the "Send" button to send a photo. To receive a photo, open the gallery where you want to receive the photo and click the "Receive" button.

You can also send photos via email. To do this, ensure you have signed into the google account you want to use on that device. Once that is done, follow these steps to send your pictures to another person via email.

- Open your gallery and choose the photo you want to send. Once you select your chosen photo, locate the hamburger button (the 3 dots at the top right side of your screen) and click the share button.
- A couple of options will open up to you. From these options, select "Gmail." This will open you up to the Gmail app and start a new conversation.
- Enter the email address to which you want to send the photo, enter an email headline (if you want), the body of the message (optional), and click the send button. A confirmatory message will pop up on your screen once your message has been sent.

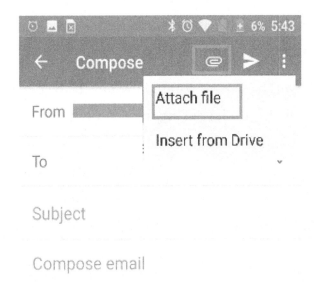

How to Set Up a Gallery

If you've never set up a gallery before, you may feel overwhelmed at the prospect of doing that for the first time. But don't worry - we're here to walk you through it step by step.

First, you'll need to choose an app for your gallery. There are many different options, so spend time researching and finding the right one. Once you've decided on a platform, sign up for an account and follow the instructions for setting up your gallery.

Next, you'll need to upload your photos. Most platforms will have an easy-to-use interface for this. Just select the photos you want to upload and follow the prompts.

Once your photos are uploaded, you'll need to create galleries. This is where you'll organize your photos into collections. Again, most platforms will have an easy-to-use interface for this. Select the photos you want to include in each gallery and give your gallery a name.

Now that your galleries are set up, it's time to share them with the world! There are a few different ways to do this:

- Send a link: Most platforms will give you a unique URL for each of your galleries. You can copy and paste this URL into an email or social media post and share it with anyone.

- Embed code: Some platforms also offer to embed code, which allows you to embed your galleries directly onto your website

Here's a unique tip for you. If you are completely new to tech and not yet sure how this works, you may want to stick with just using the default gallery on your smartphone. The upside is that all your photos and videos are automatically saved to your smartphone. You can upload them to the cloud at intervals, delete the ones you want, and modify others. You can always upgrade to using other gallery apps when you feel ready or need more functionalities that your default app doesn't afford.

Now you know how to correctly import photos onto your computer from your smartphone, export them in the proper format, send them to others, receive photos from others, and set up a gallery for easy access. Never again will you struggle with photo importing or wonder how to share your photos with others. With these tips in mind, enjoy taking and sharing photos!

Chapter 14: How to Fill Predictive Text/Deactivate/Personalize Your Typing Options

You might find autocorrect annoying if you're the type to always text with different languages. Imagine typing one word, and the autocorrect twists what you meant to type. If you find yourself in this scenario and looking forward to taking full control over your text, follow these basic steps for deactivating the auto-correct feature on your Android phone.

- Locate the setting app on your Android phone.
- Scroll down on your phone's screen, click on System, then click on languages and input. Please note that the type of Android version of your phone affects where you will see the language and input features. Some Android versions have it on the settings page.
- The next step is to click on the virtual keyboard.
- Click on Gboard or the type of keyboard on your phone's device. You should be directed to your phone's keyboard setting when you click on it.

- After you have been directed to the next section, click on Text correction.
- In part for "Corrections," click on auto-correction to deactivate the autocorrect feature.

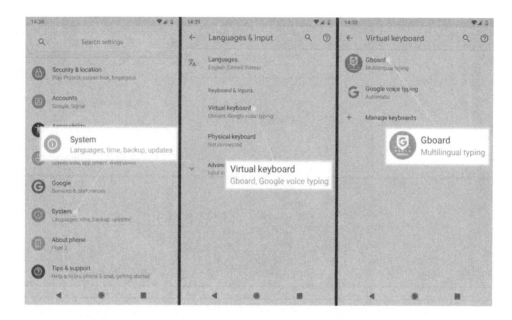

Other Types of Text Correction Features on Android Phones

The good thing about the Android keyboard is that it has been upgraded to a point where you can easily write up good and correct sentences, even when you're not a native English speaker. The feature on your keyboard can help you construct a sentence properly via predictive text on your device's keyboard. Some of the other available text correction features on an Android device are:

- Auto Space after Punctuation

One text correction feature on Android phones is the auto space after punctuation. This feature is handy if you forget to click on the space bar after making punctuation in your sentence. With this feature, you're free to type regularly without clicking on the space bar whenever you use punctuation marks like the full stop, comma, etc.

- Auto Capitalization

Another text correction feature on your Android device is the "auto space capitalization" feature. So instead of you having to capitalize the first word of your sentences, this feature will help you do that. Also, if you forget to capitalize the names of people, places, etc., this feature will help you with that too.

- Double-Space Period

The double-space feature helps to ease your stress and makes your typing faster. This feature helps you include a full stop and space by tapping on the space bar two times. However, the normal way to include a full stop and space is by first tapping on the full stop and then tapping on the space bar.

How to Deactivate Predictive Text on an Android Device

If you don't want the predictive text feature to appear on your phone's keyboard, here are some steps to deactivate it.

- Locate the setting menu on your Android device and tap on it.
- Go down your display screen and locate "system" or "languages and inputs."
- If you tap on "system," proceed to tap on "languages and inputs."
- Choose your active keyboard
- Locate the setting of your phone's keyboard app and click on it.
- Go down your display screen and tap "text correction" or something.
- Click on "predictive text" or "next word suggestion" to deactivate the feature.
- Then make use of your keyboard to ensure that the "predictive text" feature has been disabled.

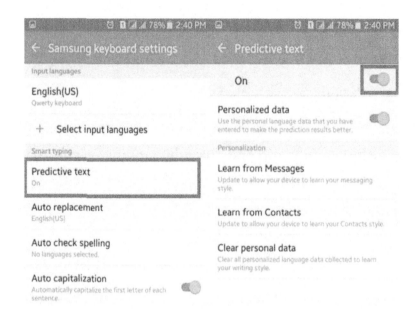

After you might have deactivated the "predictive text" feature on your Android device, your keyboard will no longer suggest text options to you while you're typing. However, the "predictive text" feature may pop up when you use apps with this feature. Also, if you don't want this feature on such apps, you must deactivate them separately. The reason is that the changes you make on your phone's keyboard do not affect the settings of individual apps.

Most people like to install other keyboards, apart from the ones that come originally with their phones. So, in that case, the active keyboard will be different. Every Android device has an integrated keyboard. However, you might install an active keyboard on your device, such as Gboard. So, if you deactivate predictive text only on Gboard, the feature will still be active on the default keyboard your phone came with. In such an instance, you must do the same process to deactivate the predictive text feature on the default keyboard.

Chapter 15: Creating App Folders

On several devices, when a new app is installed, it becomes automatically included on your home screen. Most times, it can be so tiring trying to look for the app you want to use out of the many apps you have on your phone. To make things convenient for you, group the apps into a folder to aid easy accessibility. This process requires three steps.

- Press and hold on to the app you wish to move into a folder. (that is, click on the app for some time until you get into edit mode).
- Employ the drag-and-drop method. Move the app by dragging it over another app you wish to group. Then both of the apps appear inside a folder.
- Click on Enter folder name and enter the name you wish to save the folder with.

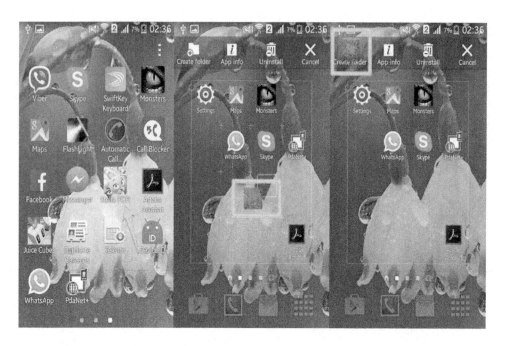

If you're planning on creating several folders, you should name them according to the apps you place in them. For instance, if you put all your game apps in one folder, you can label the folder as "Games." This process makes it simpler to find the apps.

Adding Apps into a Folder/Taking Apps out of a Folder

- Locate the bottom of the folder, and you'd see an option for you to add apps.
- Click on it and select the app or apps you wish to add to your folder.
- After selecting the apps, click on the top-right part of your phone's screen.
- Also, you can include apps in your folder from your home screen.
- You need to tap on the app you wish to move and perform the drag-and-drop process into any folder to which you wish to move the app.
- Furthermore, if you want to remove an app from a folder, click on the folder, then click and hold on the app you wish to delete.
- Locate the "remove from home" button and click it, and you are done.

How to Move Folders

Moving the location of a folder isn't a complex process. The same way you move an app is the same way you move a folder.

- Click on the folder you wish to move and hold on to it for a few seconds until you notice a slight change in your screen.
- Then, drag the folder to the new location where you want it to be. When moving a folder, the apps will shift to create space for the folder on the recent versions of Android.

Deleting App Folders

The last thing concerning folders is deleting. How can you delete a folder? This is done by removing all the apps inside the folder. Another alternative is to long press on the folder until there's a slight change on the screen. Then you drag the folder button to the remove option. This process helps to delete the folder, but your apps remain intact.

Chapter 16: How to Preserve Your Phone's Battery Life

As a new smartphone owner, you don't want your phone's battery to die suddenly, especially when needed. You can do many things to preserve your phone's battery life. Here are some of them.

Switch on Power Saving Mode on Your Device

If you find yourself in a situation where you'd need your phone to be on for a longer time, then the first thing that comes to your mind is to try to preserve your phone's battery so that it can last you for your desired duration.

One way you can do so is by turning on the power-saving mode on your device. This is to help reduce the functionality of apps that reduce battery life. For example, if you're using a Samsung device, go to your settings, then go to the battery and device care then you can proceed to click on battery entry.

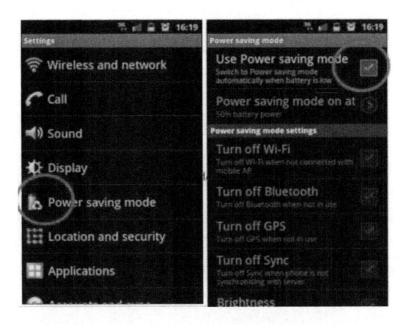

Reduce the Brightness of Your Phone

Another way you can preserve your battery life is by reducing the brightness of your device. The screens of Android devices are usually large, bright, and battery-consuming. Having your phone's brightness turned to the highest level is optional.

To do this, locate settings on your device, go to display settings, and reduce the brightness on your screen. Another alternative is to pull the notification bar and regulate the brightness level. Do this by simply swiping down from the top of your screen.

Also, if your phone's setting has been set to auto-brightness, you should disable this function. The purpose of auto brightness is to help regulate the brightness level of your phone, depending on your environment. Go to your settings and turn off the switch where you see adaptive brightness. Besides preserving your battery, doing this helps keep your eyes in good condition.

Control Your Lock Screen

A new feature was introduced for iPhone users, 'Always on Display. However, the feature has been in Android devices for some period. The purpose of this feature is to allow primary information like time and date to be seen on your phone's screen even when your device is turned off.

Even though this feature doesn't use much energy, you can still turn it off to try to preserve as much of your device's battery as possible. If you're using a Samsung Galaxy device, locate settings on your phone, look for the lock screen and turn off the switch beside Always on Display.

If you don't want to eliminate the feature, you can click on Always on Display to allow it to display only when the screen's surface is tapped.

Adjust when Your Screen Goes Off

Please note that the longer it takes before your phone's screen goes off, the more battery it consumes. Another way to preserve your battery life is by adjusting when your screen goes off. You can change this setting when your screen times out by reducing the time it takes for your device's screen to go off.

To achieve this, go to settings and locate display settings. Navigate to "screen timeout," click it, and reduce the screen timeout. Ideally, 1 minute is enough time.

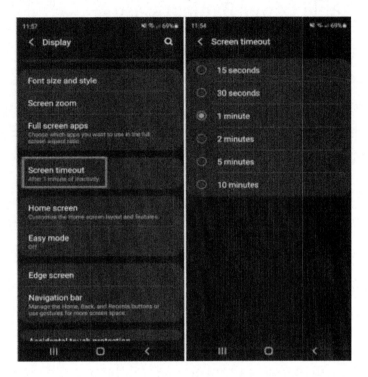

Switch off Location and Wireless Services

It doesn't matter whether you're using your phone; when so many signals are being transferred from your device, it can drastically reduce the life of your battery. If you want to optimize your phone's battery fully, there are some services in your phone that you can switch off if you're not planning on activating the power-saving mode.

For example, services like Bluetooth, wi-fi, and mobile data drain your battery with every notification you get. So, an amazingly simple way of preserving your battery life is by pulling down the notification bar and switching on the Airplane mode. Doing this will automatically sever your phone's connection to all these services and preserve its battery life.

Another backend service that drains out the battery on your device is when your phone's location is turned on. All the notifications you receive when your location service is on can reduce your battery's life. Go to settings to switch off the location services or pull down the notification bar to turn off the location from there. Although, turning off your location might hinder some other apps from working effectively.

Purchase a Portable Power Bank or Case

You can do that if you want your phone to function effectively and preserve its battery life. You need an alternative power source to charge your phone when the battery goes down. Purchase a suitable power bank that works properly with your android device. Another alternative is purchasing a battery case compatible with your specific device. However, ensure you have them fully charged before you leave your home.

Preserve Your Android Phone Battery By Avoiding These Mistakes

If you want to prevent your battery from getting spoilt and last longer, here are some common mistakes you must intentionally avoid.

Turning off all the Background Apps at the Same Time

There is this common myth that says turning off all the tabs that are on in your device's background can help to preserve the battery life of your phone. The truth is that instead of preserving your battery life, it's reducing it. How?

Turning off the entire background apps makes it more difficult for your device to restart them again. Restarting an app requires a lot of processing and uses so much of your battery. If your apps are turned on in the background, they are in the ready-to-go mode, so they don't use much of your phone's battery.

Charging Your Phone until it Reaches 100%

Oops! You are probably guilty of this.

Everyone loves to charge their phone till it reaches 100% because it brings a different level of satisfaction that the battery is fully charged. However, charging your phone's battery until it reaches 100% puts a lot of stress on lithium-ion batteries due to the increased voltage.

Therefore, the best is to charge your device until it reaches around 80-85% and stop charging. Using your phone battery until it completely goes off also puts a lot of stress on your phone's battery. So, there are two main lessons to be learned here. Do not overcharge your phone; do not drain your battery to 0%.

Switching on Your GPS for the Whole Day

It's no longer news that many applications on your Android device require your location to be turned on every time. Even when no app requires your location to be turned on, you can unconsciously keep it on. Please note having your GPS turned on at all times drains your battery severely. Why is this so? The reason is that the GPS on your device is actively involved in getting signals from different satellites to be on point with your precise location. So, it's best to keep your device's location turned off except when you require it to be turned on.

Overheating

Most of the time, we are always on our phones. We may use our phones to the point where our hands become warm from the heat being exuded. And this heat generated from your phone harms the life of your phone's battery. It drains it and can cause your phone's battery to get spoiled. Another instance that can cause your device to overheat is placing it under intensely hot temperatures.

Either by placing it directly under the sun or in an extremely hot environment. These instances can cause your phone to overheat and drain out your device's battery life. Therefore, reduce the amount of time you spend using your phone. Also, don't place your phone under direct sunlight or in places of extreme temperatures.

Your Phone's Brightness

Some people love to turn up their screen's brightness level to the point where it's almost blinding. Apart from causing harm to your eyes, you're also causing harm to your phone's battery. If you don't know, screens use the greatest battery on your phone. This means that the higher your screen's brightness is, the more the battery is utilized. The best thing is to keep your brightness level to a point where you don't have to squint your eyes when using your phone.

Chapter 17: Maintenance: How to Clean Your Android Phone When It Gets Dirty

Did you know you can clean your android phone when it gets dirty? Cleaning doesn't mean you should dump your phone in a bucket full of water and use the hardest sponge you can lay your hands on to scrub it. Cleaning your android phone helps you retain its original shine and can even get you excited about using it.

Follow these simple steps to clean your android phone any day.

- The first step you need to do before cleaning your Android phone is to remove your phone's case, if any and shut down your device.
- Get a microfiber cloth and gently clean the outer part of your phone. This helps to remove stains and deposits of filth. The cloth and the glass surface of your phone produce friction to aid in the removal of germs. It doesn't necessarily kill the germs but helps remove them from your phone's surface. Please note it's best to use a microfiber cloth instead of a washcloth or towel because the microfiber cloth has a larger surface area for absorbing dirt and filth on your phone's surface.

- The next step is to get a Lysol cleansing wipe. There's no cause for alarm in using Lysol wipes to cleanse your phone because they have been considered safe for cleaning the exterior of devices. The use of Lysol wipes is to help get rid of excess germs left. Also, if the wipe contains too much moisture, squeeze it out first. Then you can proceed to clean the surfaces on your phone gently. However, be careful not to touch the ports.

- Allow your phone to dry for like 5 minutes by placing it in an open space. Like every other cleansing wipe, Lysol spray improves if it's allowed to air dry on the cleansed surface for about 5 to 10 minutes.

- With a neat paper towel or a microfiber cloth, clean off any remaining moisture on your phone's surface. Please ensure you don't use the same microfiber cloth you used in step 2, but reach out for another one that has been washed with a sanitizer.

- The last step is for you to clean your phone's case. The same process you used to clean your phone's exterior surface should also be used to clean your phone's case. However, you can use more intense cleaners when cleaning your phone's case. This is because several phone cases are produced from exceptionally durable plastics.

Chapter 18: Android Phone Warranties (How to Know if You are Eligible for It and Get After-Purchase Support for Your Phone)

Each device you purchase comes with its warranty, so the task is up to you to inquire about the warranty of the device you're purchasing. However, some rules are common to most Android phones as regards their warranties. Often, your device manufacturer will promise to be responsible for every mechanical and electrical fault that can hinder your phone from working effectively. If you purchase a phone with a damaged screen or encounter battery problems while using your phone, then you're eligible to ask for a replacement.

For instance, you can purchase an Android device and get a three-year warranty. If your device develops a problem within those years, the manufacturer will be eligible to repair it. However, this may not happen when you're responsible for the damages, or you have annulled your device's warranty in a way.

What Your Android Phone Warranty Does Not Cover

Please note that phone warranties cannot be likened to phone insurance. This means that if you're the cause of the issue your phone is experiencing, then you are likely not eligible for a replacement. For instance, if you mistakenly drop your phone in water or accidentally drop your phone on the floor, then the manufacturer will not be responsible for any damage to your phone.

Also, if you have tampered with your device or tried to make some repairs on the phone yourself, it means you have rendered your phone's warranty invalid. Therefore, do not attempt to repair your phone during your warranty. Phone warranties also don't cover theft. So, don't walk into your phone dealership and request a new device when the previous one gets stolen unless you are willing to swipe your card through the checkout point.

Always read the warranty manual with your phone to know its terms and conditions every time you get a new device.

How to Get After-Purchase Support for Your Android Phone

The best thing to do when you purchase a new Android device is to keep the purchase receipt and the warranty document safe. Doing this helps prevent any hassle when you claim a warranty. So, if you need to claim a warranty, you either reach out to your retailer or the phone's manufacturer to gather information about what needs to be done to become eligible for the warranty. In most cases, you'd be told to reach out to a service center and return the phone.

You might encounter unfortunate situations where the retailer tells you that he's not responsible for the warranty and that you should contact the manufacturer. In some worst cases, the manufacturer will also try to pass you off to the retailer. However, in normal circumstances, the retailer has to be responsible for the warranty of your device. If they demand a receipt, and you don't have one, you can use a credit card statement showing evidence of purchase.

What are You Responsible for?

You might be responsible for the postage fee when sending your phone back to the manufacturer or retailer. Also, you might stay without a phone when you send the device to the manufacturer (except if you have a backup device somewhere else). The manufacturer will conduct some tests to determine whether they're responsible for the device's fault or you are. So, if they find out the phone's fault is yours, they'll send you the repair cost. However, you can go ahead with the repair or decline it. But if you don't want them to repair your phone, they'd charge you for the test carried out and return your device the same way you sent it.

Service and handling costs are normal.

Before you send your phone to the manufacturer, ensure you back up your important files and remove your sim/memory cards if there are any in your device. Also, ensure you have deactivated any security software that can hinder them from accessing your phone. The best idea is to get rid of all the data on your phone before sending it for the sake of privacy.

Also, note that your exact phone might not be returned to you. Manufacturers can change some parts of the device with newer parts. Sometimes, they might send you another phone of the same type and version.

Chapter 19: Best Tips and Tricks with Android Devices

Android devices come with an enormously powerful operating system. Android phones are packed with top-notch features and extraordinary tricks that help to save time. It's time to have first-hand knowledge of the awesome tricks that are on your Android device. You might have proper knowledge of how to browse on your phone or make calls, but how well do you know your device? Here are some tips and tricks you can use on your Android device.

Activate Developer Mode

The developer's mode is a popularly known trick available on Android devices. However, some people are still in the dark about this feature. Here's how it works.

- Locate the settings app on your Android device

- Move up your display screen until you see "About', then click on your device's build seven times.
- Then you'd get a countdown on your screen.
- Then you'd be sent a congratulatory message. You've now become a new developer on your Android device.

Turning this on is not compulsory, as you can maximize your smartphone even with developer mode off. Before turning this feature on, you may want to talk to an android phone expert and have them examine your phone model so they can give you their candid advice.

If you are completely new to the "android" and "tech" blocks, consider leaving this feature alone until you have figured out its relevance to your phone.

Alter the Animation's Swiftness

The latest versions of Android devices are being made to feel so swift. But some lower versions of Android devices come with 4GB of RAM, and in severe cases, they come with 2GB. This can be solved by altering your device's window animation scale, Transition animation scale, and Animator duration scale. You can change it from a speed of 1 to 0.5. Please note that changing your device's animation speed doesn't make it faster but makes you feel like it is fast while you're using it.

- Locate the settings app on your Android device.
- Find and tap on "System."
- On the developer's option, go down your screen a little to find the required options. Please note that before applying this trick on your device, you must first have activated developers mode.

Disallow App Defaults

It can be frustrating when you're going through a particular app, and links start opening to other websites instead of in the browser. Perhaps it's a link to a video on YouTube, Instagram, or Twitter - you'd then have to wait until your device cuts off chrome or any browser you're using.

There's an easy solution to this problem.

- Locate the setting app on your device.
- Scroll down to Apps and notification and tap on it.
- Look for the app that's always opening.
- Once you've seen the app, click on Advanced.
- Move up your display screen to open by default and click on it.
- Then click on clear defaults.
- Change your Wi-Fi rapidly.

Yes, how to rapidly change between the Wi-Fi network you're using to another is not obvious. Many people are used to going to settings, clicking on network and internet, and then changing the Wi-Fi network. But there is a quicker method of switching between Wi-Fi networks.

- Swipe down your display screen from the top of your device. The settings menu will be displayed.
- Do not tap on the Wi-Fi icon because that only turns it off. Instead, hold down on the Wi-Fi icon for a few seconds.
- After this, a list of the available networks in your area will pop up on your screen. Select the one you want to connect to, and you're in. it's that easy!
- Use the quick settings menu without swiping down.

Many people are used to swiping on their phone's screen to access the quick settings menu. However, you don't need to go through the stress of double swiping before accessing the menu.

All you need to do is swipe down your phone's display screen with your two fingers simultaneously.

Screen Casting

Do you have a Chromecast on your Android device? If yes, you can conveniently mirror your device's screen on another device by activating the cast feature.

- Swipe down from the top of your display screen to access the quick settings menu.

- Tap on screencast
- Then your device will locate your Chromecast

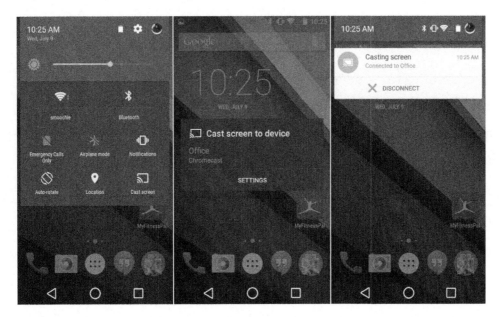

Identify Disturbing Notifications

You might get annoying app notifications from time to time, and you cannot stop them because you are unsure which app is causing them. You can resolve the issue by simply holding the notification for a few seconds, and your device will provide you with information about the source. You can then navigate to the app and disable notifications for it.

Control App Notifications

Another cool trick on Android devices is controlling notifications from apps.

- Locate the app that you want to control the notifications
- Press on the app for a few seconds
- An information icon will pop up on your screen.
- Click on the icon, and notification settings for the app will be displayed.
- You can either block all notifications from the app or some of them. You can also make it dominate priority mode or keep confidential information.
- Activate DND (Do not disturb mode)

Do not Disturb is an underappreciated and underutilized feature on Android devices. DND mode allows you to keep your device silent while allowing some sounds to pass through. The sounds could be from WhatsApp messages, app notifications, or alarms. How do you enable do not disturb mode on an Android device?

- Locate settings on your Android device
- Scroll up your display screen and tap on sounds
- Tap on Do Not Disturb
- Then choose the options that you need.

A quicker way to activate the do not disturb mode is by switching it on and off from the quick settings menu.

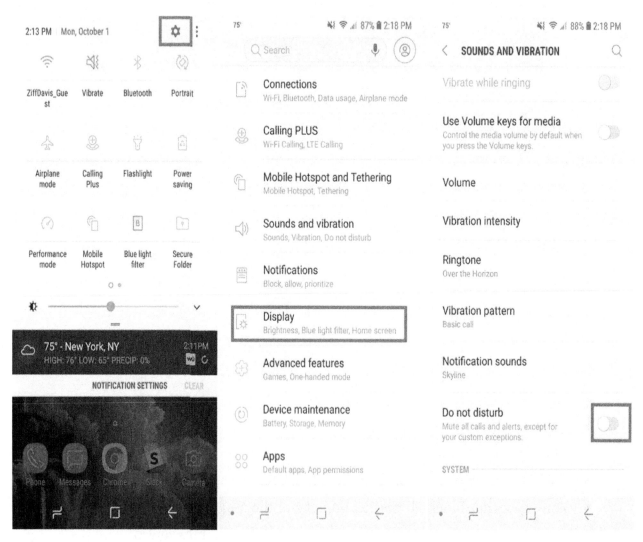

Map Zooming

This trick is handy, especially when using Google maps while driving or one-handed.

All you have to do is open your Google Maps app, double-tap the screen to enlarge it, and double-tap again to return it to its original size. In the standard step, people pinch their two fingers together on the screen while holding the phone with the other hand.

To have a more sharply defined control, tap on the screen two times, then keep your fingers on the screen. You can move your finger up or down if you need to zoom in and out.

Activate Your Device's Notification History

We would have mistakenly wiped away our notifications at one point or the other. It can be annoying, especially when you weren't privileged to check the app where the notification came from. For Android versions above 11, a solution to this issue can be fixed in the settings app on your phone. What do you do?

- Locate the settings app on your Android device.
- Scroll up on your display screen and tap on Apps and notifications.
- Then click on Notification history and activate the button beside the option.
- After activating the button, click on the notification history to reflect on what you missed.

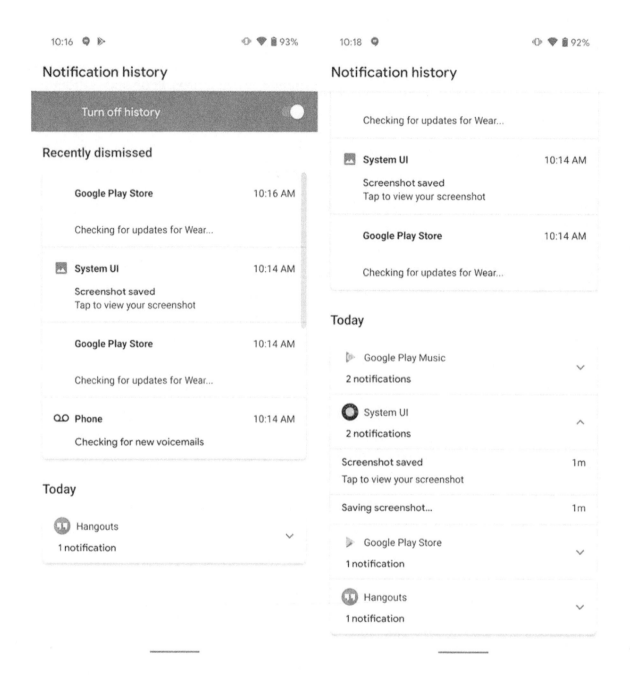

Smart Lock

Most people prefer to lock their phones with patterns, passwords, and pins for privacy. However, typing in your password or pin every single time can be annoying when you're in a safe environment or at home. At such times, a smart lock can be especially useful.

Smart lock uses GPS-specific networks, face ID, and even your voice to keep your phone from locking. Therefore, you have easy accessibility to your phone without much hassle. How can you set up a smart lock on your Android device?

- Locate the settings app on your phone
- Scroll up your display screen, then tap on Security.
- Then tap on the smart lock to activate the feature.

Screen 1 — Settings

SETTINGS

Connections
Wi-Fi, Bluetooth, Data usage, Flight mode

Sounds and vibration
Sounds, Vibration, Do not disturb

Notifications
Block, allow, prioritize

Display
Brightness, Blue light filter, Home screen

Wallpapers and themes
Wallpapers, Themes, Icons

Advanced features
Games, One-handed mode

Device maintenance
Battery, Storage, Memory

Apps
Default apps, App permissions

Lock screen and security
Lock screen, Face Recognition, Fingerprints, Iris

Cloud and accounts
Samsung Cloud, Backup and restore, Smart Switch

Google
Google settings

Accessibility
Vision, Hearing, Dexterity and interaction

Screen 2 — Lock screen and security

LOCK SCREEN AND SECURITY

PHONE SECURITY

Screen lock type
Pattern, Fingerprints, Face

Face Recognition
Your face has been registered

Fingerprint Scanner
2 fingerprints have been added

Smart Lock
Unlock your mobile ... cally when trusted locations or
other devices hav ...

... such as Auto lock and Lock instantly with

LOCK SCREEN AND ALWAYS ON DISPLAY

Always On Display
Show the Home button and a clock or information on the
standby screen

Information and FaceWidgets
Select what to show on the Lock screen and Always On Display

Notifications
On

App shortcuts
Select apps to open from the Lock screen

SECURITY

96

Conclusion

Android phones are one of the most common phone types people use nowadays. Also, they are affordable and common. The best part is that you don't need to be tech-savvy before you can effectively operate an Android device. Most Android phones come with product manuals that can help you navigate your way around the phone.

From the manual, you can get information on the basic features present on the phone, the important parts of the phone, how to carry out certain functions, and so on. With a user guide, you'd know how to set up your phone without much hassle properly and learn how to use the basic applications on your phone. You'd learn to care for your phone properly to increase its durability and physical appearance. You'd learn about the hidden tips and tricks on your device. You'd be fully informed on how to preserve your phone's battery life.

This book was created to be a supplement to your user guide. We've covered everything from setting up and optimizing your new Android to taking pictures and making videos. You've also learned about your device's warranty and how to claim if one is required.

Android phones are life savers. Imagine having the whole world in the palm of your hands simply because you have purchased a power bank and mastered how to use it for your good. There's no stopping you now. Use the information you have gathered from this guidebook to remain on top of your game as a new android user.

Then again, don't forget to keep exploring. Once you know your smartphone's basic features, feel free to tinker around. Who knows what other fun features you will discover?

Printed in Great Britain
by Amazon

25307802R00057